U0157963

粤雅小丛书

Elegant Guangdong
Series

Maolong Brush

茅龙笔

笔墨新韵

南方日报出版社
NANFANG DAILY PRESS

·广州·

粤雅小丛书编委会
编

图书在版编目（CIP）数据

茅龙笔：笔墨新韵 / 粤雅小丛书编委会编. — 广州：南方日报出版社，2020.12

（粤雅小丛书）

ISBN 978-7-5491-2262-2

Ⅰ. ①茅… Ⅱ. ①粤… Ⅲ. ①写字毛笔－制造－介绍－新会区 Ⅳ. ①TS951.11

中国版本图书馆 CIP 数据核字 (2020) 第 221583 号

MAOLONGBI：BIMO-XINYUN

茅龙笔：笔墨新韵

编　　者：	粤雅小丛书编委会
出版发行：	南方日报出版社
地　　址：	广州市广州大道中 289 号
出 版 人：	周山丹
责任编辑：	阮清钰　陈　静　黄敏虹
翻　　译：	吴忠岫
责任技编：	王　兰
责任校对：	魏智宏
内文设计：	邓晓童
封面设计：	书窗设计
经　　销：	全国新华书店
印　　刷：	广州市岭美文化科技有限公司
开　　本：	787mm×1092mm　1/32
印　　张：	2.125
字　　数：	30 千字
版　　次：	2020 年 12 月第 1 版
印　　次：	2020 年 12 月第 1 次印刷
定　　价：	25.00 元

投稿热线：(020) 87360640　　读者热线：(020) 87363865

发现印装质量问题，影响阅读，请与承印厂联系调换。

目录

【引言】

　　明代成化年间，著名的理学家、教育家、诗人、书法家陈献章在广东新会的圭峰山上筑茅舍讲学。相传一日有客人来访，临行向陈献章讨一幅墨宝。陈献章手头刚好没有好用的毛笔，灵机一动，拔了山间盛产的茅草，"束茅为笔"，蘸墨挥毫，居然笔画刚硬有力，飞白生动活泼，欣喜之余陈献章将其称作茅龙笔。因世人称陈献章为"陈白沙先生"，故茅龙笔又称"白沙茅龙笔"。

　　其实茅龙笔制作讲究，远非民间故事里说的那样简单。但有两点说得没错，一是茅龙笔为陈献章所创，二是茅龙笔以茅草制成，且尤以新会圭峰山生长的茅草为佳。

数百年过去，起源于明代、兴盛于清代、民国的茅龙笔制作技艺，在20世纪30年代抗日战争爆发后已经失传。现在生产的茅龙笔制作技艺，是后人拆解流落民间的茅龙笔，琢磨陈献章书法作品特色，一步步反推出来的。2008年，白沙茅龙笔因制作技艺的独特性和唯一性，作为代表岭南文化的一个品牌，被收入第二批国家级非物质文化遗产名录，从此得到更好的保护和发展。

第一章

笔起草莽

—— 茅龙笔的起源

其人：巨儒白沙　奠基心学

陈献章（1428—1500），字公甫，号石斋，世称白沙先生，广东新会人。陈献章年少成名，19岁中举，入国子监读书，但两次科考皆不中，后拜理学大儒吴与弼为师，回到家乡苦读10年，38岁时受学士钱溥的邀请再次入京科考，依旧不中，便居乡讲学，开创"江门学派"，主张"学贵知疑""独立思考"的自由开放学风。

陈献章的最大贡献，是摆脱程朱理学教条，创江门学派，与后来王阳明创立的姚江学派承上启下，把心学推向高峰。他所奉行的"静坐澄心，从中悟道"的哲理，并不是禅宗

白沙先生像

那般参禅顿悟的意思，而是强调博览群书后形成一种独立思考的精神。历代书籍浩如烟海，不能一味死读死背，人云亦云，而要静下心来理解思考，形成自己的思想认识，对不懂的问题，要大胆提出疑问。"小疑则小进，大疑则大进。疑者，觉悟之机也。"这样才能真正收获知识。他不满当时奉行程朱经典的学术壁垒，主张"以自然为宗"，即打破教条因袭和思想僵化，生动活泼地学习，追求不拘泥既定模式，海阔鱼跃、天高鸟飞、自由自在的思想境界。明末清初著名学者、中国思想启蒙第一人黄宗羲赞誉陈献章的成就："有明之学，至白沙始见精微。"

陈献章是一位杰出的教育家，门下弟子达数千人，遍及全国各地。他教学不论贫富，一视同仁，不仅传授知识，还注重学生品德修养，学生包括后来身兼礼、吏、兵三部尚书的思想家、大儒湛若水，台阁首辅梁储。在他逝世后，朝廷颁发诏令建造陈白沙家祠。后万历帝又诏准其从祀孔庙，追谥文恭，陈献章成为有史以来广东从祀孔庙的唯一学者。

其制：茅出圭峰　四步四德

　　关于陈献章发明茅龙笔，弟子张诩所作的《翰林检讨白沙陈先生行状》记载："山居，笔或不给，至束茅代之。"陈献章所居的山，是新会圭峰山，那里盛产的茅草特别柔顺，极富弹性。《新会乡土志·植物》载："白茅，生于山坑中，明陈献章采以制笔，今犹效之。"

　　茅龙笔的制作分为选草、浸泡制作、锤草、刮青削草四个步骤，刮削好后需要进行浸胶和梳草，用丝线绕扎茅草而成笔杆，茅龙笔就制作好了。为了达到一支好毛笔所具备的"尖、圆、齐、健"四个要素，要把制好后的茅龙笔进行"拉锋"，即把笔泡入水中，拉扯笔锋进行调锋。

选草

浸泡

绕扎

束草

笔成

　　由于茅龙笔制作技巧依靠制笔师傅自主探索，如茅草要选择不老不嫩，浸泡凭技术控制时间，砸扁茅草根部，根据所需笔形和用笔效果用利刀进行快速刮削等，每一道工序全凭制笔师傅经验进行调整，所以每一支茅龙笔，都是独一无二的。

其笔：茅龙笔吟　震动中原

史料记载南宋理学家朱熹也曾束茅为笔，写过大字且留存于世，但是将茅草做笔并发展出一种独特书风乃至一种流派，陈献章是首创者。他不满明初以来呆板拘恭、端正匀圆的"台阁体"，意图在书法上创新，追求"法而不囿，肆而不流，拙而愈巧，刚而能柔"的奇特风格。这促使他寻找一种新的书写工具，来实现自己的书法创作理想。

白茅植物纤维制成的茅龙笔，比以动物毛制成的笔粗硬，吸墨性差，

传承至今日益精美的茅龙笔

书风却苍老枯劲、雄浑大气，生发一种意想不到的"拙而愈巧"的艺术效果，一扫明代书体丰腴秀美之态，给人以极强的视觉冲击力。

当时的主流书坛受朱熹及其理学影响颇深，书风和程朱理学一样，一笔一画，法度严格，不能有半点随性。陈献章用茅龙笔所书写的茅龙书法拙中带巧、刚中带柔，生动活泼，彰显了书法个性，无疑具有划时代的革命性意义。世人耳目，为之一新，后世以"白沙震动中原"誉之。

第二章

超圣入神

——茅龙书的艺术

书风：峭削槎枒　自成一家

　　陈献章在书法上追求"动上求静"，他的茅龙书法具有狂放不羁的特点，运笔极具动感。屈大均称赞他的茅龙书法："奇气千万丈，峭削槎枒，自成一家。" 麦华三在《岭南书法丛谭》中评价陈献章茅龙书法，则用了"惊蛇投水""渴骥奔泉"来形容，道出陈献章茅龙书法动感十足的一面。

　　陈献章的书法理论与他的心学思想融为一体，不拘泥于法度，撤百家藩篱，追求自得意趣。他在书法艺术中真正达到了宇宙在我，心游八极，率意而为。

惊蛇投水般的茅龙笔牵丝与飞白

　　屈大均说陈献章的茅龙书法："熙熙穆穆，岂非超圣入神，而手与笔皆丧者乎！""熙熙穆穆"即一种既庄重又欣喜的审美形态，意思是陈献章的茅龙书法已达到一种"超圣入神"的境界，仿佛手和笔都已经不复存在，书法是自然而然形成的，需要鉴赏者用心去体会。

诗赞：茅君用事　入手神工

　　陈献章对自创的茅龙笔和茅龙书法是颇为自得的，把茅龙笔比作"茅君"，写诗对其进行赞美。比如以"茅君稍用事，入手称神工"，赞美茅龙笔的神奇；"手里龙蛇不可降""束茅十丈扫罗浮"则以极富浪漫主义的笔法描绘了自创茅龙书法的特征，凸显他对茅龙笔的喜爱。《送茅龙》一诗中，他将自己制作的茅龙笔赠送给友人，同时描绘了茅龙之气势，"赠尔茅根三百丈，等闲调性到千张"特别强调了茅龙书法所展现出的独特"调性"和不凡气度。

茅龙笔传承人张瑞亨在写茅笔字

　　陈献章用茅龙笔写的字在当时引领风尚，
人人都以拥有一幅他的茅龙笔墨宝为荣。他的
门下弟子，如湛若水、王渐逵、赵善鸣等都跟
随其学习茅龙书法，传其衣钵。后人对茅龙书
法也一直赞誉有加。明代王夫之赞美了茅龙书
法笔墨飞舞的自然洒脱：

　　　　白沙飞舞茅龙，一瓠埋头蝇迹。

　　　　莫道我狷彼狂，共弄暮天空碧。

　　　　——明·王夫之《书陈罗二先生诗后·其一》

张瑞亨先生的茅龙笔艺术展

清初彭孙遹还写有《陈白沙草书歌》歌咏茅龙书法：

晚年信手作大字，落笔纵横有奇致。
何必规规王右军，淋漓时复成高寄。

一百多年后，同乡诗人胡方重拾茅龙笔，临习陈白沙茅龙书法，作《白沙先生茅笔草书歌》表达对陈献章茅龙笔书的认识和仰慕：

胡市此山原巨兔，氄毛昔错当蒿莱。
拔归制作凭神明，去肤留骨渍蟫灰。
红藤束缚坚且直，管者其苗锋者荄。
葛陂竹杖龙变成，行时往往闻风雷。

陈献章的茅龙书法独具风格，而这些赞颂茅龙书法的诗作更为其增添了意味无穷的魅力，使其成为岭南文化中的亮丽瑰宝。

墨宝：《慈元庙碑》 风骨长存

陈献章用茅笔书写的石刻《种蓖麻诗卷》《大头虾说》等，均堪称茅龙书法经典。而陈献章去世前一年完成的茅龙书法作品《慈元庙碑》，因其独特的书法艺术和深刻的历史意义，被誉为"岭南第一碑"。1958年，周恩来总理在视察江门市新会区崖山古迹时，专程观看过《慈元庙碑》。

《慈元庙碑》所立的慈元庙故址在700多年前南宋最后一个皇帝赵昺的行都。宋亡国200多年后，陈献章倡议在此兴建慈元庙，纪念当时以身殉国的杨太后。庙里立《慈元庙碑》，由陈献章亲自撰文并以茅龙笔书写。碑高1.8米，宽1.05米，额书"慈元庙碑"4字为行楷，字径9厘米。正文600多字，跋文100多字，行书、草书兼之，字径3—5厘米。

茅龙笔：笔墨新韵

《慈元庙碑》碑题

《慈元庙碑》局部

《慈元庙碑》局部

茅龙笔：笔墨新韵

《慈元庙碑》局部

陈献章晚年的书风，在此碑中展现得淋漓尽致。整幅作品风格鲜明，布局严谨，与文字内容相得益彰，却又于章法中直抒情怀，自然成章，独具白沙茅龙书法之特点。彼时他的茅龙书法已炉火纯青，碑文字迹苍劲有力，枯润有致，浑然一体。粗看之，通篇纵横错落，气势雄伟，疏密结合，张弛有度；细观之，用笔矫健奔放，险劲峻峭，行云流水，古朴雄奇。

《慈元庙碑》在书法上有很高的艺术价值，也是一份重要的历史文献，与陈献章毕生以教化为己任的教育目标相通，鉴史以证得失，通过思考南宋灭亡的原因，达到正国风民风的深一层的目的。

第三章
笔儒天下
——茅龙笔的传承

发展：笔业沉浮　名扬海外

　　清代后期，茅龙笔慢慢进入上层阶级，新会在康熙年间出现专门制作、销售茅龙笔的店铺，一直到清末都未间断。有名的如"捷元斋笔庄"，后迁往香港易名为"捷元笔庄"。

　　民国以后，茅龙笔的制作得到发展，新会出现很多制作茅龙笔的店，笔市兴旺繁荣，甚至形成了具有规模性的"笔街"。抗日战争爆发后，战火连年，新会"笔街"被日军侵占，制笔业凋零，制作茅龙笔的匠人流散到香港，后来仅香港有零星几家茅龙笔行。

　　1978年中日建交后，日本代表团来华访问，日本使者谈及中华人民共和国成立前曾经

关山月用茅龙笔画的梅

托人到中国购买茅龙笔的往事，提出想买一批茅龙笔回国。茅龙笔这一岭南民间艺术瑰宝，重回人们的视野。

2011年2月，中国文化艺术协会与日本东京画院在东京联合举办了"中日书画交流精品展暨中日友好书画交流研讨"活动，江门市书法家龙钧球先生用茅龙笔书写的行草《曹操观沧海》《毛主席诗歌》两幅作品，入选"中日

今人观摩茅龙笔绘画

书法交流精品展"。他用茅龙笔书写的陈白沙
描写明代江门风土人情的诗歌"江门春浪两涯
平，半醉船儿天上行。坐冷烛花归问夜，莲蓬
津鼓欲三更"作品被日本东京画院收藏。

　　2008年6月，"白沙茅龙笔制作技艺"因
其历史价值、艺术价值、审美价值和环保价值
入选国家级非物质文化遗产名录。

传人：保护传统　推陈出新

　　白沙茅龙笔的传承人张瑞亨从事茅龙笔研制接近20年，他自己成立手工作坊制作茅龙笔，用茅龙笔写字作画推广茅龙笔。他制作的白沙茅龙笔曾获广东省首届民间工艺精品展优秀奖，被广东省博物馆和法国留尼旺国家图书展览馆收藏，岭南著名书画家高剑父、关山月、刘海粟、吴作人等也使用过茅龙笔作画。

　　张瑞亨父辈经营古旧家具买卖生意，他小时候在旧家具中发现木板上有陈白沙书法，还找到过几支破旧的茅龙笔，对茅龙笔产生了兴趣。1978年，中日建交以后，日本代表团来中国进行文化交流，提出购买茅龙笔的要求。

茅龙笔传承人张瑞亨

新会工艺美术厂成立茅龙笔试制组，研制茅龙笔。17岁的张瑞亨凭借深厚的绘画功底通过考试，成为一名制作茅龙笔的工人。

由于茅龙笔的使用者少，没有经济效益，4年后，工厂被并入制衣厂。张瑞亨为了不让茅龙笔技艺失传，把制笔工人请到家里来制作，并积极对茅龙笔进行创新。20世纪80年代，张瑞亨独创"茅龙排笔"用于国画创作，形成了茅龙皴法独特风格。有书画家认为用茅龙笔作画比写字更显特长，用茅龙笔作画的人越来越多。20世纪90年代初，张瑞亨被调进新会冈州画院当负责人后，建立手工作坊茅龙轩，用半卖半送的形式推广茅龙笔。

2007年，张瑞亨在报纸上发现邓某向中国工商总局申请"茅龙"商标的初审公告，为了不让江门地区独特的文化遗产成为别人获利的东西，他准备资料，自费委托代理机构办理了异议申请，成功终止邓某的商标申请，阻止了新会独特文化遗产被抢注。

现在，冈州画院里设立了茅龙轩制笔文化博览馆，观众可以了解茅龙笔历史，亲身参与茅龙笔制作。相关部门2014年协助建成新会石涧公园的茅龙草堂，使茅龙文化有了更多展示载体。草堂除了种植茅草，还有茅龙笔制作工艺、茅龙笔书法展示等活动，让游客切切实实感受茅龙笔的魅力。

如今，张瑞亨尝试扶持一些小的制笔作坊，鼓励他们生产高质量的茅龙笔，传承复古工艺；鼓励他们使用较为贵重的材料，如用紫

张瑞亨向青年人讲述茅龙笔文化

檀、花梨木做笔架，用玉雕或者石雕做笔头，让茅龙笔走高端路线。"我希望，通过茅龙笔在商业市场的流通和我院入选省级非物质文化遗产传承基地，茅龙文化的传承和发展能得到更多支持。"张瑞亨说。

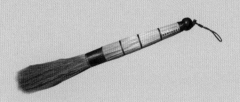

Maolong Brush:
Fresh Appeal of Writing Brush

CONTENTS

Chapter III

The World of Calligraphy—Inheritance of Maolong Brush Making / **017**

Foreword

During the Chenghua period of the Ming Dynasty, Chen Xianzhang, a famous neo-Confucianist, educator, poet, and calligrapher, built a cottage to educate his disciples and spread his ideas on the Guifeng Mountain in Xinhui, Guangdong. Legend had it that, a friend visited one day and asked Chen for a piece of calligraphic work. It happened that, Chen Xianzhang just did not have any handy brushes available, and then he hit upon an idea. He plucked some cogongrass which abounded in the mountains, "bound such into a brush". He dipped it in ink and penned the work with a flourish of the brush. The strokes were bold and rigorous, the unfilled space vivid and lively. Overjoyed, Chen named it the Maolong brush. As Chen was styled by the contemporary as "Master Chen Baisha", the Maolong brush is also called "Baisha Maolong brush".

Actually, making Maolong brush is pretty sophisticated, far from the simplicity as told in folklores. However, there were two points to the point in the folklore, one is that it was invented by Chen Xianzhang, the other is that the brush is made of cogongrass, and preferably that grows on the Guifeng Mountain in Xinhui.

Hundreds of years have elapsed, Maolong brush making technique, which originated in the Ming dynasty and flourished in the Qing Dynasty and the Republic of China, had been lost after the outbreak of the War of Resistance against Japanese Aggression in the 1930s. The Maolong brush that is now made is the result of step-by-step reverse engineering by dismantling whatever could be collected of Maolong brush from private collectors and having the characteristics of the calligraphic works by Chen Xianzhang analyzed. In 2008, the Baisha Maolong brush was placed in the catalog of the second batch of national intangible cultural heritage due to the unique and singular craftsmanship in its making and as a representative brand of Lingnan culture. It has since got better protection and undergone better development.

Chapter I

Humble Origin of Maolong Brush

The Person: Baisha, a Towering Intellectual and Founder of Philosophy of the Mind

Chen Xianzhang (1428—1500), byname Gongfu, self-styled Shizhai, known as Master Baisha by his contemporary, was born in Xinhui, Guangdong. Chen attained fame while he was young. He passed the provincial civil service examination and was admitted to the Imperial College at the age of 19, but failed twice in the imperial palace examination. Later, he went back to his hometown to study was Wu Yubi, a master neo-Confucianist for 10 years. When he turned 38 years old, on the invitation of the grand secretary of the Imperial Academy Qian Pu, he went to the imperial capital again and participated in the imperial palace examination, however he failed a third time. He finally settled in his hometown and gave lectures, founded the "Jiangmen School", advocating free and open intellectual atmosphere of "valuing the inquisitive spirit in learning" and "independent thinking".

Chen Xianzhang's biggest contribution comes from his shaking off the doctrines of the neo-Confucianism represented by Cheng Brothers (Cheng Hao and Cheng Yi) and Zhu Xi. The Jiangmen School formed a connecting link for later Yaojiang School founded by Wang Yangming, the two jointly brought the philosophy of the mind to its peak. Unlike Zen's practicing meditation as a means to gaining enlightenment, the philosophy of "sit quietly, clear the mind, enlighten yourself" he upheld stresses forging an independent speculative spirit after extensive reading. On the backdrop that books handed down from past generations amount to staggering quantity, one cannot rely on mechanical reading and rote learning, or parroting others' ideas. Instead, one should calm the mind to think on one's own and form one's own understanding, and put forward questions without hesitation when coming across any difficulty. "Small question brings forth small progress, big question big progress. Question is the key to enlightenment." Only in this way can one attain true knowledge. He was dissatisfied with the academic barriers created by the canons of the Cheng Brothers and Zhu xi which were upheld by his contemporary. He proposed to "follow the law of nature", that is, to break away from the doctrines and conventions, to free the ossified mind; one should learn rigorously with a lively spirit, seek knowledge without following slavishly established norms, pursue a realm of the mind where it is only bound by the unfathomable sky and sea, set free the mind. Huang Zongxi, a well-known scholar at the closing period of the Ming Dynasty

Cogongrass, the raw material of Maolong brush

and the beginning of the Qing Dynasty, the first enlightenment thinker in China, lavished praise on the achievements of Chen Xianzhang, "the learning of the Ming Dynasty takes a turn for the subtlety and essence in the hands of Baisha."

Chen Xianzhang was an outstanding educator, had thousands of followers from across the country under his tutorship. He treated all disciples equally regardless of their wealth. He did not only impart knowledge but paid more attention to the cultivation of the character of his disciples. His disciples included Liang Chu who served as the grand secretary of the secretariat, and Zhan Ruoshui who was a great Confucian scholar and later headed three ministries, namely, the ministries of Rites, Personnel and War. After his death, the imperial government decreed that a memorial place be built, named Baisha Clan Memorial Hall. In the Wanli period, the emperor further decreed that Chen be enshrined and admitted to the Confucian temple to receive accompanying sacrifice. He was posthumously awarded the title Wen Gong. He was the only scholar from Guangdong admitted to the Confucian temple receiving accompanying sacrifice.

The Making: Made with Locally Grown Grass, Stringent Steps and Virtues

Regarding the invention of Maolong brush by Chen Xianzhang, his disciple Zhang Xu recorded in *A Biographical Sketch of Master Chen Baisha—A Historiographer of the Imperial Academy*, "lived in the mountain, occasionally there was a shortage of brush supply, the master was forced to bind cogongrass in place of brush." The mountain wherein Chen Xianzhang lived is the Guifeng Mountain in Xinhui, the cogongrass which abounds in the place is particularly pliable and extremely resilient. The Chapter "Plants" of *The Local Annals of Xinhui* has the following entry, "cogongrass grows in pits of mountains, Chen Xianzhang of the Ming Dynasty had plucked it to make it into brush, this practice is still followed to date."

The making of Maolong brush consists of four parts, namely, selecting grass, soaking, hammering, removing green peel and knifing grass. The knifed grass has to be gummed and combed, and then the grass straws are bound with silk thread into a handle, a Maolong brush is made by now. In order to make a brush with the four factors "sharp, round, even, stiff" with which a fine brush should be complete, the brush so made is then subjected to "tipping", that is, dipping the brush into water and adjusting the brush tip by stretching it.

Each Maolong brush is unique in its making, for instance, in the entire making process, the cogongrass chosen must be neither to old nor too tender, control of the duration for soaking grass, hammering flat the grass roots, quick knifing according to the requirements for brush shape and writing effect, each process requires the brush-making craftsman to make appropriate adjustment as his experience dictates, while such technique and skill can only come from instruction and personal examples received as well as individual exploration by the craftsman.

007

The Brush: Maolong Brush Calligraphy Rocked the Literary Circle

Historical records have entries that the neo-Confucianist Zhu Xi of the South Song Dynasty had bound grass into writing brush and written big characters therewith which has been handed down to our time. However, making grass into writing brush, developing such into a unique writing style, and even founding a school actually originated with Chen Xianzhang. Dissatisfied with the "Secretariat Style" which featured woodenness, primness, upright and roundness prevalent among scholars since the establishment of the Ming Dynasty, Chen sought to innovate calligraphy writing, pursuing an extraordinary style which was "conforming to principles while breaking out of the box, unrestrained without going to unscrupulous, appearing clumsy but all the more artful, rigid without losing pliability". This prompted him to find a new writing instrument to realize his ideal of calligraphic creation.

Maolong brush made of cocongrass fiber is thicker and stiffer than brush made of animal hair, inferior in ink absorption capacity. However, its writing style is hoary, rugged yet rigorous, magnificent and majestic, generating an artistic effect of "appearing clumsy but all the more artful", which cleared the plump and graceful style of the Ming Dynasty calligraphy and brought the viewer a powerful visual impact.

The mainstream contemporary circle of calligraphy was under the heavy influence of Zhu Xi and the neo−Confucianism thereof; the prevailing writing style was exactly like that of the neo−Confucianist like the Cheng Brothers and Zhu Xi, strict adherence to the established writing regime without a gleam of spontaneity in each and every stroke penned. The Maolong brush calligraphy by Chen Xianzhang featured clumsiness without expelling ingenuity, rigid with softness, vivid and lively, manifested the personality of the calligraphy. It was indeed of epoch-making significance, refreshed the mind of the contemporary people. Later generations sang the praise of him with the statement "Baisha Rocked the Country".

Chapter II

Transcending Worldliness—The Art of
Maolong Brush Calligraphy

Writing Style: Rugged Branches Styled a New Calligraphy School

Chen Xianzhang pursued a style of "seeking immobility from mobility", the Maolong brush calligraphy of him featured wild and unrestrained flourish, the flow of strokes permeated with great momentum. Qu Dajun commended his Maolong brush calligraphy as, "striking force shooting up tens of thousand feet, as though stiff branches shooting forth, unique in his own style." Mai Huasan in *Essays on Calligraphy of Lingnan Area* commented on Chen Xianzhang's Maolong brush calligraphy as "that the startled snake shooting into water" "thirsty steed galloping to a fountain", which delineated the extremely dynamic aspect of Chen's calligraphy.

In the pursuit of contentment and life interest, Chen's theory of calligraphy fully blended with his thought on philosophy of the mind, which shatters the obstacles of established traditions and shakes off restrictions set up by the myriad predecessors. He attained in his calligraphic art the real mental realm of holding the entire universe in the self, perusing the eight extremities of the universe with the mind, acting spontaneously on the spur of the moment.

Qu Dajun described the Maolong calligraphy of Chen Xianzhang as "vigorous and majestic, is it not that it transcends the realm of the sainthood, while the hand and the brush both seem lost to him!" "Vigorous and majestic" refers to an aesthetic state of both stateliness and ecstasy, which means Chen's Maolong calligraphy attained a state of "transcending the sainthood" where both the hand and the brush appear nonexistent, as though the calligraphy comes into being naturally; it requires the reader to apply his mind and appreciate the beauty of it.

Poem of Laud: Maolong Brush Consecrating Handwork

Chen Xianzhang took great pride in the Maolong brush and Maolong calligraphy he invented, he likened the Maolong brush to "gentleman Maolong", and composed a poem to laud its virtue. For instance, he sang the praise of the wonder-working property of Maolong brush with the statement of "slight effort by gentleman Maolong consummates the work of hand". "The python and dragon in the hand could not be rendered docile", while words like "I'd bind cocongrass into ten-meter sweep and clean the Luofu mountain" described the characteristics of Maolong calligraphy invented thereby in romantic depiction, highlighted his love for Maolong brush. In the poem titled *Gifting Maolong Brush*, he described gifting his friend with handmade Maolong brush and the majestic bearing of Maolong, "Gifting you with grass-root a hundred meter long, in leisure time you may edify the mind by writing calligraphies a thousand strong." He laid special stress on the unique "tone" and extraordinary majesty of Maolong calligraphy.

012

Chen Xianzhang's Maolong calligraphy set the writing trend of the time, and his contemporary took pride in having a piece of his Maolong calligraphy. Many of his disciples including Zhan Ruoshui, Wang Jiankui and Zhao Shanming studied Maolong calligraphy under his guidance and passed on his mantle. Later generations heaved praises on Maolong calligraphy. Wang Fuzhi of the Ming Dynasty acclaimed the naturalness and unrestrained easy-going of the flourishing Maolong calligraphy:

Baisha wields a Maolong brush, a gourd buries in fly trails.
Say not I am conceited and he presumptuous, together we luxuriate under the blue sky in dusk.
Wang Fuzhi of the Ming dynasty (appeared in The First of the Series of *Afterthoughts on a Letter to Masters Chen and Lou*)

In the early Qing Dynasty, Peng Sunyu also wrote *Ode to Chen Baisha's Cursive Calligraphy* singing the praise of Maolong calligraphy:

In his later years, he casually wrote large characters, varied strokes bring exquisite appeal.
Why bother to follow slavishly Wang Xizhi, in hearty rendition he again attains lofty ideal.

More than a hundred years later, fellow poet Hu Fang brought Maolong brush back to life from neglect, imitated and learned from Chen Baisha's Maolong calligraphy. He wrote *An Ode to Master Baisha's Maolong Cursive Calligraphy* to express his understanding and admiration of Chen Xianzhang's Maolong brush calligraphy:

The mountain of the savage people used to grow giant rabbits, its fine hair mistaken for grass.
With divine prowess it is captured, with skin removed and bone retained to be impregnated with clam shell ash.
Bound with red vine it is rigid and straight, the stalk is made into the handle and the root the tip.
The dragon transformed into the dragon—bamboo cane, in moving it brings forth roar of the wind and thunder.

Chen Xianzhang's Maolong calligraphy is unique in its style, and these poems acclaiming Maolong calligraphy introduced infinite charm to it, making it a brilliant gem in the Lingnan culture.

Treasured Calligraphy: The Most Revered Stele of Lingnan

Stone inscriptions including *Scrolls of Poems on Growing Castor* and *On Big Head Shrimp* written with Maolong brush by Chen Xianzhang could all be justifiably listed among classics of Maolong calligraphy. The Maolong calligraphy work Ciyuan Temple Stele completed one year before Chen Xianzhang's death is reputed as "the No.1 stele of Lingnan" because of its unique artistic effect of calligraphy and profound historical significance. In 1958, Premier Zhou Enlai paid a special trip to visit the "Ciyuan Temple Stele" when inspecting the Yashan Historic Site in Xinhui District, Jiangmen City.

The former site of the Ciyuan Temple wherein the Ciyuan Temple Stele was seated was the temporary capital of Zhao Bing, the last emperor of the South Song dynasty more than 700 years ago. More than 200 years after the fall of the Song Dynasty, Chen Xianzhang proposed to build the Ciyuan Temple in the site to commemorate the emperor's mother who sacrificed her life for the country at that time. The *Ciyuan Temple Stele* was erected in the temple, the content of which was written in Maolong brush by Chen Xianzhang himself. The Stele is 1.8 meters high and 1.05 meters wide. The 4 characters (three words in English) on the stele head, Ciyuan Temple Stele are written in regular script, with a diameter of 9 centimeters each. The body text is more than 600 characters long, and the postscript more than 100 characters long, running script and cursive script are alternately employed, with a diameter of 3—5 centimeters.

Chen Xianzhang's calligraphy style in his later years is vividly displayed in this stele. The whole work has a distinctive style of

Maolong brush soaked with ink

its own, which is rigorously laid out; the layout and content of the text highlight and complement each other. It expresses unadorned sentiment while conforming to the bounds of composition, forms an independent entity without affectation, manifests the unique characteristics of Baisha Maolong calligraphy. At that time, his Maolong calligraphy had attained consummation, the inscriptions on the stele were vigorous and powerful, charming with appropriate rigidity and plumpness, forming an integral entity. At a glance, the whole article appears crisscrossing and interlacing, with majestic momentum, properly spaced out and with appropriate intensity; at closer inspection, the strokes are vigorous and unrestrained, bold and precarious, with flowing elegance, simple and majestic.

The *Ciyuan Temple Stele* attains high artistic value in calligraphy and is also an important historical document. The stele accords with Chen Xianzhang's lifelong educational mission of enlightenment. By learning from the history to assess gains and losses, examining the reasons for the demise of the South Song Dynasty, he revealed his further attempt at rectifying the prevailing national mentality and folk morals.

Chapter III

The World of Calligraphy—Inheritance of Maolong Brush Making

Development: Ups and Downs of the Brush-making Industry

In the late Qing Dynasty, Maolong brush slowly found its inroad into the upper class. During the Kangxi period, shops specializing in making and selling Maolong brushes were opened in Xinhui, which stayed in business uninterrupted until the end of the Qing Dynasty. Famous ones included "Jieyuan Zhai Brush Store", which later moved to Hong Kong and renamed "Jieyuan Brush Store".

After the founding of the Republic of China, the industry of Maolong brush gained some development, and there came into being many shops making Maolong brushes in Xinhui. Maolong brush business prospered, and "a brush-making street" of substantial scale had taken shape on Huimin Road in Xinhui. After the outbreak of the War of Resistance against Japanese

Aggression, war raged on for years; the "brush-making street" of Xinhui was occupied by the Japanese troops, and the brush-making industry declined. Craftsmen of Maolong brush-making scattered and exiled to Hong Kong. Later, there remained only a few Maolong brush shops in Hong Kong.

After the establishment of diplomatic relations between China and Japan in 1978, a Japanese delegation visited China. The Japanese delegation talked about past stories on having entrusted someone to buy Maolong brushes prior to the founding of the People's Republic of China, they proposed to buy a batch of Maolong brushes and bring them back to Japan. By then, Maolong Brush, a gem of Lingnan folk art, has returned to people's vision.

In February 2011, Chinese Culture and Art Association and Tokyo Painting Academy jointly organized the "Sino-Japan Calligraphy and Painting Masterpiece Exhibition and Sino-Japan Friendship Calligraphy and Painting Exchange Seminar" in Tokyo. Long Junqiu, a calligrapher from Jiangmen City, had two pieces of work penned with Maolong brushes in running script admitted to "Sino-Japan Calligraphy and Painting Masterpiece Exhibition", titled *Cao Cao Viewing the Grand Sea* and *Chairman Mao's Poem* respectively. While his calligraphy work written with Maolong brush was collected by Tokyo Painting Academy of Japan; the work is a poem composed by Chen Baishi, describing the Jiangmen customs of the Ming Dynasty, "Spring waves leveled the two banks of Jiangmen, half intoxicated I rode a boat in the sky. Seat grew cold and candle snuff flickered, I asked for the time of the night, the Lianpeng Ford was about to sound the third watch."

In June 2008, "Baisha Maolong Brush-Making Technique" was placed in the Catalog of National Intangible Cultural Heritage for its historical, artistic, aesthetic and environmental protection values.

Inheritor: Protecting Tradition and Innovating

Zhang Ruiheng, the inheritor of Baisha Maolong brush-making technique, has been engaged in developing and making Maolong brushes for nearly 20 years. He has set up a handcraft workshop to make Maolong brush, worked calligraphy and painting with Maolong brush to promote the brush. Baisha Maolong brushes made by Zhang won Excellence Award of the First Folk Crafts Exhibition of Guangdong Province, and was collected by the Guangdong Provincial Museum and the French Reunion National Book Exhibition. Famous Lingnan calligraphers and painters including Gao Jianfu, Guan Shanyue, Liu Haisu, Wu Zuoren have also painted with Maolong brushes.

Zhang Ruiheng's father ran the business of buying and selling antique furniture. When he was a child, Zhang found books of Chen Baisha's calligraphy on the wooden boards in old furniture. He also found several worn Maolong brushes, which aroused Zhang Ruiheng's interest in Maolong brushes. In 1978, after the establishment of diplomatic relations between China and Japan, a Japanese delegation visited China on cultural exchanges and requested to purchase Maolong brushes. Xinhui Industrial Art Factory set up a Maolong brush trial development group to develop Maolong brush. The then 17-year-old Zhang Ruiheng passed the screening exam with his solid drawing skills and became a worker in making Maolong brush.

Due to the scarcity of users for Maolong brush and hence little economic benefits arising therefrom, the factory was merged into a garment factory four years later. In order to prevent the loss of Maolong brush-making skills, Zhang Ruiheng invited brush-makers to his home to do the making, and conducted active innovation on Maolong brush. In the 1980s, Zhang Ruiheng created his unique "Maolong Row-Brush" for the creative work of Chinese paintings, formed a unique style of Maolong brush's cun (a method of diluting the ink for lighter shades or texture) method. Some calligraphers and painters think that painting with Maolong brush can better highlight the painter's strength, and more and more people begin to use Maolong brush for painting. In the early 1990s, Zhang Ruiheng was transferred to Xinhui Gangzhou Painting Academy in charge of the academy. He then

Zhang Ruiheng, the inheritor of Baisha Maolong brush making technique

established a handicraft workshop Maolong Xuan, engaged in promoting Maolong brush in half-sale and half-present manner.

In 2007, Zhang Ruiheng came into a notice of first action in newspaper which said that a certain Mr. Deng had applied for the initial review of "Maolong" trademark to the State Administration for Industry and Commerce. In order to prevent the unique cultural heritage of Jiangmen from becoming a means for profit to others, he prepared materials and entrusted an agency to handle it at his own expense. The opposition application successfully terminated Deng's trademark application and prevented Xinhui's unique cultural heritage from being registered unjustifiably.

Presently Maolong Xuan Brush-Making Culture Museum has been set up in Gangzhou Art Academy. Visitors can learn about the history of Maolong brush, participate in the making of Maolong brush. Pertinent departments assisted in the completion of the Maolong Thatched Cottage in Xinhui Shijian Park in 2014, which brings forth more carriers for displaying Maolong culture. In addition to planting cocongrass, there are also scheduled activities such as Maolong brush-making craftsmanship and Maolong brush calligraphy on exhibit, allowing visitors to get a firm grip on the charm of Maolong brushes.

Nowadays, Zhang Ruiheng tries to support some small pen-making workshops and encourages them to produce high-quality Maolong brushes. Inheriting retro craftsmanship, using more precious materials, such as red sandalwood and rosewood for pen holders, jade or stone sculpture for brush sleeve in an effort to bring Maolong brush upscale. "I hope that through market transaction activities and our academy's acceptance into the provincial base for intangible cultural heritage inheritance, the inheritance and development of Maolong culture will win more support." Zhang said.